Brief Encounters in the Sky

15 Chilling True Stories of the Unexplained That Will Have You Believing We Are Not Alone!

Chase Carter

Table of Contents

UFOs – The Phenomena that will Not Go Away

Interest in Unidentified Flying Objects, or UFOs, has been high since the 1940s and it is easy to see why. There have been dozens of sightings of these phenomena by credible individuals.

Fighter pilots, Air Force officers, airline pilots, and even Air Force generals have gone on record as saying they've seen UFOs. Fighter pilots have fired on the objects, pilots have made emergency landings to avoid them, and generals have dispatched military units to investigate them.

More importantly, many of these people have stated that the objects they saw displayed capabilities beyond those of earthly technology. Some witnesses have described objects that did things that seemed to violate the laws of physics as we presently know them.

Despite these claims from credible individuals, the claims have been ignored, attacked, and sometimes suppressed by the government and the media. It isn't clear why the government and media want to suppress UFO reports, but the claims of witnesses have often been met with attacks. These attacks include efforts to dismiss all claims, even credible ones, as optical illusions or cases of mistaken identity.

No matter how they are greeted, UFO reports may document a legitimate phenomenon that should be investigated and taken seriously, particularly if we want to answer the question, "is there is something or somebody living in the universe besides us?"

1976 Tehran UFO Incident

One of the most intriguing UFO incidents of all time occurred over the Iranian capital of Tehran in 1976. The incident is fascinating because it involved an object that displayed the capability of interfering with and shutting down the weapons systems on fighter jets.

Strangely enough, we know a great deal of the details of the incident because the Iranian government shared everything it knew with its close ally, the United States. In 1976, Iran was ruled by the Shah, or Emperor Mohammad Reza Shah Pahlavi. The Shah's government received a tremendous amount of support, including military assistance, from the United States. The Imperial Iranian Air Force, which worked closely with the U.S., turned all of the data it had on the incident over to the U.S. military.

That means we actually know more about the Tehran incident than many UFO encounters in the United States. The reason for this is that the Tehran investigation was not conducted through normal channels, so the information was not classified and covered up.

Lights in the Sky

Ironically enough, most of the witnesses never actually got a good look at the object involved in the Tehran incident. Instead, many people spotted a very bright light in the sky over the city on Sept. 19, 1976. The light was so bright that somebody eventually called General Nader Yousefi, the air force commander in the area. Yousefi didn't believe the accounts until he looked out the window and saw the light in the sky.

As soon as he saw the light in the sky, which was described as bright as a star, the general scrambled to a F-4 Phantom II fighter jet. The object in the sky was so bright that the jet's pilot, Capt. Azizkhani, could see it from 70 miles away.

When the captain flew towards the object, something very disturbing happened. All of the electronics in his plane stopped working, yet the plane itself remained in the air, so the Captain simply turned around and went back to his base. Azizkhani later stated that he did pick up a radar signature of something as large as a Boeing 707.

Two UFOs in One

When a second F-4 took off and approached the UFO, something even more unusual happened. A smaller object flew out of the UFO and came straight at the second F-4. The pilot of that plane, Lt. Jafari, believed that he was under attack and tried to launch a Sidewinder missile, but all of his electronics, including his weapons, suddenly shut down.

The object followed Jafari for a short period of time and then returned to the larger UFO. At the same time, Jafari saw another object leave the UFO and land on the ground near the Rey Oil Refinery on the outskirts of Tehran. It isn't clear what landed, but it was bright enough to light up the area on the ground.

The electronic interference was not just confined to the jets. The personnel in the control tower at the Mehrabad International Airport reported that many of their systems shut down. The crew of a civilian airliner that was coming for a landing at Mehrabad also reported a communications failure. Unlike the fighter pilots, the civilian pilots saw the UFO and reported a cylinder-shaped object with bright lights on each end and a flashing light in the middle.

The Landing Site

The next day, Sept. 20, 1976, the airmen who had seen the UFO took a helicopter out to see if they could find the object that had landed. They didn't see anything, but their radio picked up a strange signal described as a beep coming from a small house nearby.

The people in the house reported that they had heard a loud noise and seen a bright light like lightning the night before. The witnesses didn't see anybody or anything else.

An investigation of the landing site was carried out, and this inquiry reportedly included radiation testing. It is not known who carried out the investigation or testing, nor has any documentation about the testing ever been located. It isn't even known if the tests were done by American or Iranian authorities.

The reason for this is that the Shah's government was overthrown three years later in the Iranian Revolution. The results of the testing might have been lost or they might be in the hands of the Revolutionary Guard. Since the Revolutionary Guard is vehemently anti-American, it is doubtful that they would share those documents.

Interest at the Highest Levels

The Tehran incident sparked interest at the highest levels of the U.S. government. A four-page Defense Intelligence Agency (DIA) report on the encounter was distributed to President Gerald Ford, Secretary of State Henry Kissinger, the Joint Chiefs of Staff, the National Security Agency, and the Central Intelligence Agency. It is safe to assume that the report was also distributed to foreign leaders, such as the Prime Minister of Great Britain and the Shah.

In the DIA report, the Iranian officers involved stated that they did not believe the object sited was of terrestrial origin. Another Iranian general, Abdollah Azarbarzin, later told the TV show *Sightings* that he thinks aliens were searching for a way to contact people on Earth. Azabarzin stated that no country in the world had technology capable of creating the interference with electronics that the pilots and control tower reported on Sept. 19, 1976. The U.S. General in charge of Air Force operations in Iran in 1976, Richard Secord, refused to talk about the incident publicly.

The fact that the President and other top officials were briefed indicates that the U.S. government was taking UFOs seriously and believed that they might have been a threat. It is also interesting to note that the Iranians were far more willing to discuss their real beliefs in public than the Americans were.

The Tehran incident bears some similarities to other UFO sightings, including the notorious Battle of Los Angeles in 1942 and U.S. Air Force Lt. Milton Torres's encounter with a UFO over England in 1957. Like the Iranian pilots, Torres reported that his fighter jet's weapons were shut down by a mysterious force when he tried to fire on a UFO. The Battle of Los Angeles involved bright lights in the sky over the city.

Henry Kissinger and the UFOs

There was at least one other UFO sighting on Sept. 19, 1976, near Rabat, Morocco, and that incident involved an unusual tube putting out mysterious lights. The object attracted so much attention that the U.S. Embassy in Morocco was asked about it.

Interestingly enough Henry Kissinger, who had seen the reports on the Tehran incident, took the trouble to respond to the request from Morocco. Kissinger sent a memo stating that all UFOs could be explained by natural causes. His statement contradicts the findings of the U.S. Air Force, which has found that there have been a number of UFO incidents that cannot be explained.

The Tehran incident is perhaps the most credible UFO sighting of all time. Two Air Force generals saw it and stated publicly they believed it was of alien origin. Their belief was shared by an experienced air traffic controller who was on duty in the tower. Unlike many UFO claims, the Tehran incident cannot be readily dismissed or explained.

2007 Alderney UFO Sighting

The objects seen over the English Channel on April 23, 2007 included one of the largest UFOs ever sighted. One of the witnesses, Capt. Ray Bowyer, an airline pilot, estimated that the object might have been a mile in diameter. The sighting was of interest because there were reports that the two UFOs were picked up on radar.

Bowyer sighted the object when he and another pilot were flying a small airliner from Southampton in Southern England to the English Channel island of Alderney, which is one of a number of islands off the coast of France that are part of Great Britain. Bowyer had been making the flight for eight years when he saw the objects.

There had been some sightings of unexplained lights in the sky in the same region 10 weeks earlier. Bowyer was serving as a co-pilot when he spotted the object around 3 p.m. At first, Bowyer thought he saw a reflection of light, but when he examined the object with his binoculars, he realized that it was solid. When he looked again, Bowyer realized that there were two objects.

Other Witnesses

Radar operators on the island of Jersey picked up two objects that moved slowly southward for about 55 minutes. The operators noted that whatever they were picking up lacked transponders to identify them. All aircraft over a certain size must carry such transponders.

A number of people, including tourists on the island of Sark and at least one other pilot, observed unusual objects in the sky. The passengers on the plane also observed the phenomena.

Surprisingly, one of the other witnesses was another airline pilot who was flying in the area, Capt. Patrick Patterson. He was flying from the Isle of Man to Jersey.

The Ministry of Defense and the French military denied that they had any aircraft in the area. Experts noted that the radar used could not have accidently picked up the radar signature or the reflection of a ship or a ferry, as some debunkers have said. It isn't clear if the objects were picked up by military radar.

Capt. Bowyer later said that he had learned that the military was aware of the objects, but was not overly concerned about them. Bowyer didn't say how he had learned that the military knew about the objects. The objects might have been picked up by military radar or surveillance satellites.

The Objects Grew Bigger

Whatever objects Capt. Bowyer saw were definitely very large. In fact, the objects were so large that Bowyer underestimated their size. He first though they might be the size of a Boeing 737 airliner, but he later realized they were larger. He described the objects as bright blue and giving off a bright light.

Bowyer's plane continued on its normal flight and set down on Alderney. During the approach to Alderney Bowyer spotted the second object, which he described as smaller than the first.

Still Unexplained

The Alderney UFO sighting is still unexplained, although the most popular explanation is doubtful. Some people believe that Bowyer might have spotted a secret military aircraft, but it does not seem possible that the existence of craft the size of the ones that Bowyer saw could be kept secret.

Some observers noted that Bowyer's description resembled earlier UFO sightings because he claimed that the craft was saucer shaped. His descriptions resembled craft seen by an aircrew near Paris in 1994. That sighting involved a giant craft which was also picked up by French military radar. Interestingly enough, that craft simply disappeared after being sighted for a period of time.

Other explanations that have been advanced include an optical illusion created by an earthquake. There had been a quake near Folkestone, England a few days earlier. It isn't clear how an earthquake could have created such an illusion in the air.

Media Attention

The Alderney UFO sighting, unlike most modern UFO reports, attracted a great deal of media attention, largely because Bowyer was willing to talk to the press. His descriptions intrigued TV interviewers.

Bowyer and a number of other UFO witnesses spoke at the National Press Club in Washington D.C. on Nov. 12, 2007. He was one of several pilots who discussed encounters with UFOs and asked for more investigation into the matter.

The size of the craft Bowyer reported indicates objects that it would not be possible to create or fly with modern technology. There is no way to power craft that size or keep them in the air with the technology presently available on this planet. It is clear that Ray Bowyer saw something that was beyond the capabilities of modern aeronautical engineering.

Carson Sink UFO Incident

The Carson Sink UFO Incident was one of several encounters between U.S. military aircraft and UFOs that took place in the 1950s. It is also considered a "good" or credible UFO sighting by researcher and author Edward Ruppelt.

The incident is considered a good sighting because of who reported it: two colonels in the United States Air Force. The witnesses were Lt. Colonel John L. McGinn and Lt. Colonel John R. Barton. The two were both stationed at the Pentagon at the time of the incident and were familiar with the latest aircraft technology. Interestingly, the nature of their true jobs at the Pentagon was classified.

Encounter over Nevada

The incident began with a routine flight on July 24, 1952. The two colonels took off from an airfield near San Francisco in a B-25 bomber; they were heading for Colorado Springs, Colo.

The B-25 was passing over the Carson Sink, the area of desert north of Reno, Nev., roughly the route of modern Interstate 80, when they had their encounter. Around 3:40 p.m. one of the men noticed three unusual wedge-shaped aircraft. At first they assumed the vehicles were F-86 fighter jets, but they noticed the craft were flying in a tight V formation. The objects were also flying far too high to be F-86 jets.

When they estimated the speed of the craft, they realized they were flying faster than any known jet at the time. The objects also had no tails or flight canopies, which meant they were not airplanes. The three craft then flew to within a few hundred yards of the bomber before flying off.

When Barton and McGinn finally landed in Colorado Springs, they called Air Defense Headquarters and learned that no known military or civilian aircraft had been flying over the Carson Sink at the time of incident. That report itself is suspect because we know of one military plane over the Carson Sink at that time: the colonels' B-25 bomber.

A Very Detailed Report

The Carson Sink incident is unusual because the Colonels made a very detailed report about it. The experienced pilots noted that the UFOs were flying too high for regulations and that they flew at three times the speed of an F-86, the standard U.S. Air Force fighter jet in 1952.

Investigators from Project Blue Book, the Air Force's official enquiry into UFOs, looked into the matter. The investigators noted that the colonels described the craft as delta winged and similar to jets then used by the Navy. A call to the Navy revealed that they had no such planes in the area and that most of the naval aircraft in the region were painted bright blue.

Interestingly enough, both colonels had been publically skeptical of stories about "flying saucers" before the Carson Sink incident. After the encounter, they were both ready to believe in UFOs.

Ruppelt and other experts are likely to believe the report because the two men who filed it were familiar with almost every airplane being flown in the world at the time. They would have been familiar with the planes flown by the Soviet Red Air Force and America's allies, such as the British Royal Air Force.

Project Blue Book also did a very thorough investigation of the incident, probably because of the colonels' political connections. The investigators determined that there were no balloons or naval aircraft in the area. They also looked into the possibility of experimental military aircraft and couldn't find any evidence of that either.

A Good UFO Sighting

The Carson Sink incident is a good UFO sighting because two extremely experienced pilots with hundreds of flying hours reported it. The incident is intriguing because the two men involved were extremely familiar with military aircraft and with the performance of aircraft.

We can conclude that the two colonels saw some sort of flying craft and that those flying craft exhibited capabilities that greatly exceeded those of any known to exist in 1952. From their description we also know that the objects were manufactured and under the control of an intelligence capable of complex aerial maneuvers.

Project Blue Book listed no known explanation for the Carson Sink incident. It simply left it as unknown, which means the Air Force has no explanation for what the colonels saw.

Encounters such as the Carson Sink Incident, the Mantrell Incident, and the Gorman Dog Fight demonstrate that the U.S. military had a number of encounters with UFOs in the 1940s and 1950s. These encounters also demonstrate that the pilots involved believed that they had seen manned craft that were produced by a technology far superior to any on Earth.

Yet the Air Force stopped investigating such claims, possibly because it could find no answers. The unidentified vehicles officers like Colonels McGinn and Barton saw were too fast and maneuverable even for the hotshot pilots of the U.S. Air Force to catch up with.

Chiles-Whitted Eastern Airlines UFO Encounter

1948 was the year that UFOs, or flying saucers as they were called in the 1940s, entered the public imagination. It's easy to see why, as that year saw several notable sightings of unidentified flying objects, including the Mantell incident and the Chiles-Whitted close encounter.

The Chiles-Whitted encounter is notable and highly credible for several reasons. First, there were the two main witnesses: Clarence Chiles and John Whitted. They were both highly experienced pilots working for Eastern Airlines and decorated World War II veterans. Second, their description sounds very realistic and unlike most UFO stories. Finally, there were a number of observations of similar objects around the world at the same time.

An Eventful Flight

The encounter began with a routine flight from Mobile to Montgomery, Alabama, on July 24. Chiles and Whitted were piloting a DC-3 when they noticed something very unusual.

An odd-shaped object came at their plane at a very high speed. At first the two thought it was an experimental jet being tested by the U.S. Air Force. Yet when the object got closer, they realized that it was no Earthly aircraft because of its size; it was three times the size of a B-29 bomber and around 100 feet in length—larger and longer than any plane in the world at that time.

The object was clearly a craft of some sort, but it was completely smooth and had no wings. It did put out a red exhaust, but it made no noise and it moved so fast that it almost knocked the DC-3 out of the sky. The object gave off a vast amount of turbulence.

Intriguingly, the two pilots gave a very detailed description of what they had seen. They noted that the object had a "radar pole," a windshield, and at least three rows of windows. Both pilots insisted that the object was clearly man made.

Most of the passengers on the plane didn't see anything because they were asleep. One, Clarence L. McKelvie, woke up and said that he had seen a very bright light outside the plane but little else.

Official Investigation

Chiles and Whitted were soon the objects of media attention and an official investigation. Air Force intelligence officers working with Project Sign, an official enquiry into UFOs, interviewed them in Birmingham, Ala.

Sign investigators soon uncovered some intriguing information from witnesses. A ground crew chief at Robbins Air Force Base in Georgia stated he had seen a cylindrical object two or three times larger than a B-29 with a stream of fire coming out of its tail end. An Air Force pilot flying over the Appalachian Mountains near the North Carolina and Virginia state line said he saw a "bright shooting star" moving in the direction of Montgomery.

The most fascinating report came from half a world away in the Hague, Netherlands. Several witnesses in that city saw a UFO they described as a rocket with two rows of windows along the side. That sighting was reported on July 21, 1948, three days before the Eastern Airlines encounter. The ship described by the Dutch witnesses was similar to that seen over Alabama.

No Explanation

The incident ignited a minor controversy within the Pentagon that eventually led to the demise of Project Sign. The Sign investigators believed the pilot's stories, yet they also knew that there was no known technology that could explain the giant craft. Eventually they concluded that the object was powered by nuclear power.

The Sign investigators favored an extraterrestrial origin for the object because they could find nothing like it mentioned in technical journals. Among other things they noted that it was impossible to steer a rocket with the technology available on Earth in 1948.

A document (or Estimate of the Situation) presenting this thesis was written and forwarded to higher authorities. Yet the brass rejected the document and ordered it destroyed. The generals apparently rejected the hypothesis because there was no physical proof for it.

Cover Up at the Highest Levels

Shortly afterwards Project Sign was shut down and replaced with Project Grudge, which was dedicated to "debunking" or disproving UFO stories rather than investigating them. The Air Force also presented a new explanation that the pilots had seen a meteor. This explanation was rejected by the one astronomer to look into the incident, Dr. J. Allen Hynek, a consultant to Sign; Hynek insisted that no astronomical explanation was remotely possible.

The Project Sign personnel continued to stand by their alien hypothesis even when ordered to disavow it. To skeptics it looked like the Air Force was attempting to cover up the incident.

The Air Force brass might have wanted to avoid stoking the fires of Cold War paranoia, which was already running rampant in the United States and starting to destroy people with anti-Communist hysteria. The generals might have also been afraid of telling the American people that there were flying craft they couldn't shoot down; after all, the incident occurred less than seven years after Pearl Harbor. Americans were still very nervous about an attack from the skies.

So what did Chiles and Whitted see that night in the summer of 1948? Nobody knows; there is still no modern technology that can explain the giant craft they described. Interestingly, NASA is looking into the possibility of using low energy nuclear reactions (popularly known as Cold Fusion) as a power source for rockets and aircraft. Yet any such craft is still a purely theoretical notion today in 2013, 65 years after the incident. The extraterrestrial hypothesis still seems to be the best explanation, even though there is no proof.

Coyame UFO Crash

The 1947 Roswell incident may not have been the only time the U.S. military retrieved a crashed UFO. Some UFO researchers believe that U.S. and Mexican military forces were able to recover a flying saucer that crashed near Coyame in the Mexican state of Chihuahua on Aug. 24, 1974.

The stories of the crash are intriguing because they indicate that Mexican soldiers might have been killed in the incident. The incident also involved the crash of a small private plane that was destroyed. It is not clear what happened to the pilot of the private plane, although he or she might have been killed in the crash.

A Strange Collision

The Coyame incident began with a strange radar reading on Aug. 24, 1974. Radar operators at an unknown U.S. military base picked up a strange object in the sky that was moving at speeds of up to 4,000 miles per hour. The object crossed the Gulf of Mexico and came ashore near Corpus Christi, Tex.

At some point the object apparently collided with a Cessna 180, a small private plane that was flying between El Paso, Tex. and Mexico City. The Mexican and U.S. militaries both decided to dispatch forces to the area to see what happened.

Four U.S. helicopters, probably from Fort Bliss near El Paso, arrived at the crash site near Coyame sometime after the collision. They discovered that they were not the first ones there. Coyame is about 30 miles off the border crossing on the Rio Grande River at Presido, Tex.

Weird Objects on the Ground

When the helicopters landed, the GIs discovered a horrifying sight – four dead Mexican soldiers in a jeep. The soldiers had apparently been asphyxiated by some unknown force. The Americans also discovered the burnt-out wreckage of the Cessna and an unusual object.

The object was a metallic disc of an unknown size. The disc wasn't very big because the soldiers were able to pick it up and haul it away with a Huey helicopter. Rumor has it that the object was later taken to Wright Patterson Air Force Base in Ohio for examination. Wright Patterson has traditionally been the center of Air Force investigations into UFOs. Other reports claim that the disk might have been taken to the Centers for Disease Control (or CDC) in Atlanta for study because the military believed it contained deadly germs.

The reports about the incident are conflicting and they point to a cover-up by both the U.S. and Mexican militaries. It isn't clear what was being covered up, but the governments of both nations seemed to be involved.

Cover Up or Conflict?

Some accounts of the Coyame incident indicate that the U.S. forces engaged in an illegal covert operation in Mexico, yet the Americans were able to operate openly in Mexico without interference from the Mexican military. That would indicate that the Mexican and U.S. forces were cooperating.

U.S. helicopters were able to fly over Mexican airspace without interference, and reports indicate that a U.S. Army convoy entered Mexico to retrieve the disc. Since no Mexican citizens were present at the crash site, it is obvious that the Mexican Army was keeping them away.

There is also the question of what killed the Mexican soldiers discovered by the U.S. troops. The GIs reported that the Mexicans seemed to have been drawing their weapons when they were killed. What were they firing or aiming at? Did they provoke some sort of response from somebody or something? Some reports indicate that the Mexican soldiers might have been trying to move the disc.

More Questions

Both the Mexican and U.S. governments have denied that the Coyame incident took place. The Mexican Army denied that any of its soldiers were killed, and the U.S. government denied that anything was recovered.

Reports also indicate that U.S. fighter planes might have been scrambled to intercept the disk. It isn't known if Mexican planes were scrambled or not, nor is it known what knocked the disk out of the sky. Was it downed by the collision with the Cessna or was the disk shot down by U.S. or Mexican air defenses? It might have fallen from the sky onto the Cessna.

The object recovered has been described as a silver disk with no portholes or markings of any sort. It was around 16 feet in diameter and five inches wide. That could indicate that it was some sort of drone.

Strange Revelations

Unlike the Roswell incident, the Coyome crash was kept secret until 1991, when a letter detailing the incident was mailed to a number of UFO researchers in the United States and Europe. Those who received the letter believe it was mailed by government employees with security clearance and knowledge of UFOs. It isn't clear if the letter was mailed by U.S. or Mexican government personnel.

The mysterious letters are all that is known about this incident. No witnesses have come forward, and the mysterious disk has never been located. Its fate and present location are still unknown.

Nor is it known what happened to the Mexican soldiers who died or what killed them. Some of the reports indicate that the military believes they were killed by a biological weapon of some sort.

So what happened near Coyame on Aug. 24, 1974? Nobody knows, or if somebody knows, they are keeping it secret. It could have been a crash by some sort of satellite or experimental aircraft. One possibility is that it could have been some sort of Soviet missile from Cuba. Another is that it was a secret U.S. aircraft based on alien technology. What is known is that neither government involved in the affair wants to admit that it occurred or that anybody died.

Ellsworth UFO Sighting

What has become known as the Ellsworth UFO Sighting is an intriguing encounter because the object reported was actually picked up on radar by the U.S. Air Force. The existence of the object was verified by observers on the ground, radar, and fighter pilots who were scrambling to intercept it.

The incident is called the Ellsworth Sighting because it involved radar operators and fighter pilots at Ellsworth Air Force Base near Bismarck, North Dakota. The incident itself actually took place near the small town of Black Hawk just outside of Bismarck.

UFO on Radar

The sighting began when a woman named Miss Kellian called the base to report mysterious objects in the sky on August 11, 1953. Kellian, whose first name is not available, was part of the Ground Observer Corps, an organization of civilians that watched the skies for enemy planes. Kellian told the warrant officer who answered the phone that there was an extremely bright light in the sky.

The warrant officer asked the radar operator to scan the area around Black Hawk. The radar picked up an object in the exact location Kellian had said it would be. The object was moving south at a high speed. The airmen realized that the object would soon be over the base, so two of them went outside to take a look.

They saw a bluish-white light moving through the sky in the direction of Rapid City, South Dakota. The controllers at the base then radioed the pilot of an F-84 fighter plane that was in the air at the same time. The pilot saw the object too and decided to give chase.

Chasing the UFO

When the F-84 got close to the UFO, it began behaving like a manned craft, picking up speed and moving away. Every time the fighter jet got close to the UFO, the object moved away.

Eventually the plane started to run low on fuel, so the pilot turned around and headed home. When he did, the UFO actually followed him back to the base.

Another pilot at the base decided that he wanted to see a flying saucer, so he took off in another F-84. The pilot saw the light, but when he got within three miles of it, the objects picked up speed and moved off. At one point the second pilot turned off the lights in his plane to make sure he wasn't seeing a reflection. After a few minutes the second pilot broke off the intercept and returned to base.

The second pilot also picked up the object on his plane's radar. The object obviously was not an optical illusion; it was real.

Some reports also state that the object was seen by the crew of an Air Force C-24 transport plane that was flying over the area. The C-24's crew reported that the object was blinking green and red. Some reports also stated that there was more than one object involved.

The Chase Continues

The radar tracked the object as it moved towards Fargo, North Dakota. The Air Force personnel at Ellsworth called the Ground Observer Corps in Fargo. The Corps office in Fargo reported receiving accounts of a fast-moving, bluish-white light moving through the sky in the area. After Fargo, no more reports of the bluish-white light came in.

Capt. Edward J. Ruppelt, the director of Project Blue Book, the Air Force's official enquiry into UFOs, was so intrigued by the Ellsworth case that he personally flew to Bismarck to investigate. Ruppelt talked to the pilots and personnel involved and concluded that they had seen an object and not an optical illusion.

The case is still officially listed as unknown and unsolved. It is one of several incidents involving U.S. Air Force planes and UFOs that occurred in the late 1940s and early 1950s. It seems as if somebody was watching the U.S. military or testing its capabilities.

It's interesting to note that the Ellsworth Sighting took place in the same area as the Gorman Dogfight in 1948. In that incident, a fighter plane, a P-51 Mustang, chased a UFO near Fargo. The pilot involved in that incident, Air National Guard Lt. George Gorman, reported pursuing a ball of light.

Another Ellsworth Sighting

There was another intriguing UFO sighting reported at Ellsworth Air Force Base nearly 17 years later. In this incident, Air Force Lieutenant Michael Sehorn reported seeing a large object sitting on the ground near a silo where a Minuteman nuclear missile was housed.

Sehorn noticed the object when he investigated an alarm that went off. The lieutenant also noticed that the missile had been disabled. When the Air Police investigated they discovered that the detonator had been removed from the missile's warhead. An air policeman also reported seeing an individual climbing down the hatch to the missile but didn't get a good look at him.

UFO experts are divided on the second Ellsworth incident; at least one of them, Bob Pratt of the Mutual UFO Network, or MUFON, believes it to be a hoax. The UFO Conspiracy Website rejects this claim and displays what it states are two official Air Force documents that detail an incident at Ellsworth in 1970. Areas of the documents are blacked out, making it hard to determine what they say.

The Conspiracy Website states that the documents are an incident report from the Air Force's Office of Special Investigations. It is impossible to tell if these claims are authentic or not. If they are, it indicates that somebody was probing America's defenses and examining the nation's weapons technology.

The two sightings at Ellsworth show that curious things have happened in the sky over U.S. Air Force bases and that the organization has been investigating them. Despite the investigations, no satisfactory explanations for these sightings have been provided.

Japan Air Lines Flight 1628 Incident

The United States Federal Aviation Administration (or FAA) normally doesn't like to investigate UFOs or acknowledge that they might be a possibility. Therefore, it's quite extraordinary when the agency in charge of civil aviation in the United States investigates an encounter with an unidentified flying object.

That's what makes the Japan Air Lines Flight 1628 incident in 1986 so intriguing – the FAA actually investigated it and reported its findings to higher authorities. Even the White House was reportedly interested in the incident and the FAA's findings. The reason for the official interest was that the object was reported by a veteran Japanese aircrew.

Officials might have also been concerned because the objects were reported between two U.S. Air Force bases and only a few hundred miles from Soviet territory. The Cold War was still on in 1986, and the USSR was still very much in existence and on the minds of President Ronald Reagan and his advisors.

Fateful Transpolar Flight

The incident began with a special transpolar cargo flight from Europe to Japan. Transpolar flights hop over the North Pole between Europe and Japan. On Nov. 17, 1986, Capt. Kenju Terauchi of Japan Airlines was piloting a Boeing 747 cargo plane on a special transpolar run.

Specifically, he and his crew were hauling a cargo of French wine to Japan. It was 1986 and Japan's economy was booming, fueling the market for European luxury goods. The special flight, which included a refueling stop, took the 747 over Alaska.

Around 5:11 p.m., Terauchi and his crew were flying near Mount McKinley when they noticed two unusual craft. The craft seemed to following his plane, but he couldn't get a good look at the objects because it was already dark in Alaska in November.

Defying the Laws of Physics

Terauchi soon noticed something rather unusual – the strange craft did not seem to be affected by inertia. They seemed to defy the laws of physics and act as if gravity did not affect them. In other words, the craft displayed capabilities far beyond those of known technology.

The craft eventually reversed course and drew closer to the jet. Terauchi said they gave off so much heat that he could feel it inside his plane. He described the craft as rectangular and having glowing nozzles or thrusters. When he described the objects, Terauchi said he believed that they appeared to be under automatic control.

The objects eventually flew over the horizon and disappeared. When they did, Terauchi trained the 747's radar on the area they had come from and discovered that there was an object there. He later described an object as big as an aircraft carrier flying through the air. He said that the object also followed the 747. It is believed that this object was the mothership from which the smaller craft were launched.

FAA radar at Fairbanks, Alaska didn't pick up the object, but the radar at Elemndorf Air Force Base may have picked it up. It isn't known if the Air Force scrambled fighter jets to investigate the object or not. Air traffic control in Anchorage offered to send fighters to help Terauchi, but he declined the request.

Investigation and Interest at a High Level

Like the Tehran incident, the Flight 1628 incident sparked interest at the highest levels. The FAA did investigate and interviewed Terauchi, who stated clearly he had seen a UFO. He repeated his claims to the Japanese media, which contacted the FAA.

FAA officials didn't know what to do, although they were looking into the matter. Some of the officials involved eventually briefed Vice Admiral Donald E. Engan about the incident. Engan displayed interest, but told the officials not to talk to anybody. FAA officers said representatives of the BIB and the CIA attended the meeting and also noted that members of President Reagan's scientific study team were present.

The FAA held a press conference on March 5, 1987, in which FAA public information officer Paul Steucke confirmed that a UFO was present, yet he denied rumors that the FAA had picked up a UFO on its radar.

Punishing the Pilot

Shortly after he spoke to the press, Capt. Terauchi was removed from flight duty and transferred to a desk job. Japan Air Lines punished him for going public about a UFO by keeping him off flight duties for years afterward.

If Terauchi had not seen anything, then why was his employer so anxious to silence him? There have been allegations that pilots and others are often told not to speak about UFOs to protect the public reputation of airlines.

So what did Terauchi and his crew see, anyway? The UFO described sounds like the ones mentioned in the Tehran incident. They might have seen an experimental U.S. military aircraft, but the craft they described clearly had capabilities beyond any known military technology. Since the area of the sighting is in the area of two major U.S. Air Force installations, Eielson and Elmendorf Air Force Bases, he may have seen something monitoring U.S. military activity.

Obviously, we cannot say why a UFO would be watching our military, but we can speculate on the government's motives. Officials and military officers might not want to admit that any aircraft could operate in sensitive areas without the military being able to do anything about it. The mystery remains, and it may have damaged a man's career.

Manises UFO Incident

The Manises incident was one of the most dramatic and well-publicized UFO encounters of the 1970s. It was also one of the most credible because it involved a fighter pilot and the crew of a commercial airliner.

Unlike some incidents, the Manises encounter involved dozens of people because a loaded airliner had to change course and make an emergency landing. If that wasn't dramatic enough, a fighter plane was scrambled and missiles were fired at mysterious lights in the sky.

Drama in the Skies

The incident began peacefully enough with a
routine flight between Salzburg, Austria and
Spain's Canary Islands on Nov. 11, 1979. The
109 passengers on TAE's Super Caravelle JK-
297 jetliner wanted a peaceful holiday in the sun;
instead, what they got was drama in the skies.

The turn for the dramatic occurred nearly
halfway through the flight when pilot Francisco
Javier Lerdo de Tejada spotted a series of red
lights that were quickly approaching their airliner.
As in many nighttime UFO incidents, the crew
didn't see the object; they simply saw the
mysterious lights. Tejada was worried because
the lights were on a collision course with his
plane.

Tejada radioed the tower, but the tower had no explanation because they couldn't pick up anything on radar. A military radar in Madrid didn't pick up anything either.

It isn't known what Tejada saw, but he was so scared by it and worried for his passengers that he made an emergency landing in Manises, Spain. Tejada reported that the lights seemed to chase his plane and follow it right to the landing field.

A Fighter Plane Joins the Chase

Shortly after the airliner set down, the radar finally picked up an unknown object in the sky. Authorities were so worried that they scrambled a Mirage F-1 fighter jet from the Spanish Air Force.

The jet's pilot, Capt. Fernanda Camara, had to take his jet to Mach 1.4 (or nearly one and a half times the speed of sound) to see the object. When he hit that speed, Capt. Camara reported seeing a cone-shaped object that kept changing color. The object vanished, but ground control informed Camara that another mysterious object had appeared on radar over the nearby city of Valencia.

Camara flew to that location and fired a heat-seeking missile at the UFO. When Camara took aggressive action his jet's electronics suddenly shut down. Such interference with the electronics of warplanes had been reported in other UFO encounters, including the Tehran Incident in 1976 and a U.S. Air Force Lieutenant's pursuit of a UFO over England in 1957.

The UFO eventually flew off towards Africa, and Camara broke off the pursuit because his plane was low on fuel. The pilot returned to base with a vague description of the object.

Enquiry and Cover-up

Unlike UFO incidents in the United States and Britain, the Manises incident led to an official investigation in Parliament. Enrique Mugica, a member of the High Chamber of the Cortez or Spanish Parliament, asked for an official explanation nearly a year later in September 1980.

The explanation given by the government wasn't very convincing. The Spanish Air Force told Mugica that the pilots involved had been fooled by optical illusions. The explainers didn't say how optical illusions had been picked up on radar or interfered with the electronics on Capt. Camara's fighter plane, nor did they say what had caused the illusions in the first place.

In 1994, the Spanish Air Force's official investigation of the matter was declassified. This document provided an even more absurd explanation that flashes from a chemical factory 100 miles from Manises had created the illusion of lights moving in the air.

Further doubt is put on these theories by several witnesses who saw the red lights reported by Tejada from the ground. They reported that something was following the airliner.

What did they see?

Another explanation for the Manises incident is that the airliner and fighter plane encountered an experimental aircraft from the United States Navy's Sixth (or Mediterranean) Fleet. The Sixth Fleet was in the area of Valencia on the night of Nov. 11, 1979.

Some people have speculated that the fleet was using high-powered jamming equipment that interfered with Camara's fighter plane. It is hard to believe that the Navy would risk using such devices in areas where airliners were in the sky. It's also interesting to note that Camara didn't report interference until he took aggressive action.

The interference clearly came from the UFO and it only started when the UFO was threatened. That would indicate some sort of scanning technology that could monitor the functioning of the aircraft's electronics. The United States had no such capabilities in the 1970s, and as far as we know, still does not.

Like the Tehran incident, the Manises incident can be viewed as highly credible. As in other UFO incidents, official efforts to explain the sighting fail to hold up to examination and end up raising more doubts than answers.

Milton Torres' 1957 UFO Encounter

Milton Torres became famous as a UFO witness 50 years after he had his close encounter. In 1957 Torres, then a young fighter pilot, was ordered to shoot down an unidentified flying object over England and told to keep the incident secret or else.

More than 50 years later Torres' encounter was exposed when Her Majesty's Government declassified thousands of documents about UFOs. One of the documents contained an account of Torres' incident. That prompted the media to locate Torres, by then a retired college professor living in Miami, and ask him about the affair.

Torres not only stood by his account, but he publicly accused the U.S. and British governments of suppressing information about UFOs and what happened to him. Torres stated that he believes he encountered a manned craft of some sort.

Going after a UFO

On May 20, 1957, 25-year-old U.S. Air Force Lieutenant Milton Torres was stationed at RAF Manston, a Royal Air Force base in Kent, as part of the U.S. forces protecting Europe from the Soviets. Torres was flying the F-86D Sabre jet at the time of the incident.

Torres and another pilot were in the alert shack at Manston on the evening of the 20th, which meant they were ready to take off at a moment's notice. Late in the evening they were told to scramble or take off because an unidentified flying object had been spotted over East Anglia.

The object was picked up on radar as well as from the ground. Torres said the object behaved oddly, staying motionless for large periods of time. He also said that radar indicted the object was as large as an aircraft carrier.

Locking on a UFO

Milton Torres never actually saw the UFO he was sent after because of the thick fog over the North Sea. Instead, he saw the images of it on his radar screen. That's not unusual in modern air combat, in which missiles are the main weapons used.

The UFO was sitting at 32,000 feet in altitude, and it was moving at more than 700 miles per hour, or Mach .92, which was over the speed of the F-85D. The plane was approaching its limits and Torres still hadn't made contact.

Torres contacted the tower for instructions, and he was ordered to fire all 25 of his plane's rockets at the object. Each of the rockets contained as much explosive force as a 75 mm artillery shell. Torres followed orders and prepared to fire.

Incredible Capabilities

As soon as he had locked on, the UFO suddenly moved off at a high speed. It was out of range within seconds, so Torres turned off the missiles and returned to base. When the UFO moved off, Torres believed that it was moving at Mach 10, or 7,000 miles per hour.

Equally incredible, the UFO seemed to know when he locked on his missile. That would indicate extremely accurate sensors capable of examining the inner workings of his fighter plane.

Obviously, the craft exhibited capabilities well beyond those of 1957 technology or even modern technology. The incident was an authentic one and unexplained.

The IBM Salesman

The day after the incident Milton Torres encountered a mysterious man who he said was dressed like an "IBM salesman" in a dark blue, three-piece suit and trench coat. Torres said the man was an American who had a National Security Agency (NSA) ID card.

The man told Torres that his encounter was highly classified and that he couldn't discuss what had happened with anybody, not even his commanding officer. The man also threatened to suspend Torres' flying privileges, which would have ended his Air Force career. The man even claimed he could take Torres' pilot's license away.

Torres didn't see the man again, but he kept quiet about the incident for over 50 years until he was certain the incident had been declassified. So who was the man Torres met? That's hard to say; he might have been a CIA or NSA agent or working for a British intelligence agency such as MI-5 or MI-6 or the GCHQ (Government Communications Headquarters), the British equivalent of the National Security Agency.

Fifty-One Years of Silence

Like a good officer, Torres kept his mouth shut for 51 years and had a successful career in the Air Force. He served as a Range Control officer during the Gemini and Apollo space missions and flew 276 combat missions during the Vietnam War. Torres later taught mechanical engineering at Florida International University in Miami.

When he finally talked, Torres said the UFO he had encountered was designed by an alien intelligence and not manufactured on Earth. Torres also stated that he's glad he never fired because the aliens would have vaporized him if he had.

Milton Torres obviously encountered something in 1957. He sent an enquiry about the incident to the Ministry of Defense in 1988 but received no response. There is at least one other explanation for Torres' encounter.

The respected British newspaper, *The Times*, has speculated that Torres might have got caught up in a secret electronic warfare experiment. *The Times* noted that a CIA program called Palladium used electronic equipment to create false images on radar. Interestingly enough, Palladium did not begin until the early 1960s.

Torres could have encountered an early test of the Palladium technology or a similar experiment being conducted by the British or even Soviet governments. Yet no evidence of such testing in Britain in the late 1950s has come to light.

The truth about Torres' 1957 close encounter might be locked away in some unknown file. That means we might never know what happened on that foggy night in 1957.

Sperry UFO Case

There are some UFO sightings that can turn a hardened skeptic into a believer. That was definitely the case with Capt. Willis T. Sperry, a veteran pilot for American Airlines. Sperry was skeptical of all claims about unidentified flying objects until the night of May 29, 1950.

That evening Sperry and another pilot named Bill Gates (no relationship to the Microsoft billionaire) were flying a DC-6 airliner from Washington, D.C., to Los Angeles. The plane had just left Washington and was flying over Mount Vernon, Virginia, when the crew had an encounter with the unknown.

The pilots noticed the object because it appeared to be 25 times brighter than the brightest star. Yet it was not an astronomical phenomenon, because it was cigar-shaped rather than spherical. Sperry told investigators and the press that the object hovered and circled around his plane. This astounded Sperry because his plane was moving at 300 miles per hour at the time.

Corroborated by Other Pilots

The Sperry Case is taken very seriously by UFO experts because it was corroborated by other highly-experienced pilots. Shortly after he reported the encounter, Sperry was approached by another American Airlines pilot, Henry H. Myers.

Myers, who had been flying another airliner 400 miles south of Mount Vernon on the night of May 29, 1950, said he saw a brilliant shooting star. Myers was astounded when the shooting star changed directions and then moved horizontally through the air just like a piloted craft.

Despite corroboration and Sperry's background, the Air Force tried to ignore the incident and dismiss it as a meteor. Interestingly enough, one of the Air Force's own investigators, W.B. Klemperer, who knew Sperry, believed his account.

According to Klemperer's account, the DC-6 nearly collided with the UFO. Sperry described the UFO as looking like a submarine. There was a blue light at the end of the cigar-shaped object. This description sounds like the object encountered by Eastern Airlines pilots Clarence Chiles and John Whitted on June 24, 1950. They described a UFO as long and cigar-shaped and putting out a trail of flame.

Another Intriguing Incident

Despite the military skepticism, Sperry never retracted his claims; he kept making them in letters to *Flying Magazine* and in a 1964 interview with Los Angeles TV station KABC. During the interview, Sperry described a second mysterious incident over Moline, Illinois, in 1952.

Sperry said he was flying a DC-7 at 21,000 feet when his radio suddenly picked up static. The static was heard on every frequency, and it included noises that sounded like a high-speed record. Sperry said other pilots in the area and control tower personnel reported the same problem.

At the same time as the interference a number of pilots reported a mysterious bright light in the sky. Sperry didn't give a good description of the light, but he believes it was causing the interference with radio signals. This second object could have been an astronomical phenomenon such as a solar flare or Aurora Borealis, both of which are known to interfere with electronics.

What to Make of the Sperry Incident

So what should we make of the Sperry Incident and the earlier Chiles-Whitted encounter? These incidents obviously didn't involve known astronomical phenomenon such as meteors.

A strong possibility is that the cases involved an unknown astronomical phenomenon. That does not explain the description of a metal object similar to a submarine or the fact that the objects circled airliners.

Another explanation is that large aircraft such as airliners were a brand new development in the late 1940s and early 1950s. Some suggest that alien visitors would have been interested in the appearance of large planes in the atmosphere and might have wanted to check them out.

Official Reaction and Inaction

Sperry's recollections indicate that many airline pilots reported seeing such objects in the 1940s and 1950s but few were willing to come forward and report them or publically discuss them. The main reason for this was that pilots might have feared for their careers.

The flyers might have been afraid of being ridiculed or punished by airline executives afraid of bad publicity. When Sperry talked to the Air Force's project Blue Book UFO enquiry in 1955, he was accompanied by an Eastern Airlines public relations officer.

The accounts provided by Sperry show that pilots took UFO sightings seriously in the 1950s and discussed them openly among themselves. That indicates that there may have been more sightings like Sperry's that were not revealed to the public.

Part of the reason pilots stopped reporting such sightings was that the Air Force simply lost interest in them. Since there was no official agency recording such accounts and investigating them, people stopped reporting them. Another possibility is that whoever was piloting the UFOs simply lost their interest in airliners as they became more commonplace.

Willis T. Sperry went on to fly for many more years after his UFO sighting. In 1964, when he went on television, Sperry was flying 707 jets across the Atlantic. UFO sightings reported by such an experienced pilot should never be ignored.

The Gorman Dogfight

The incident known as the Gorman Dogfight was one of two reported clashes between UFOs and the U.S. Air National Guard over the United States in 1948. The other was the Mantell incident, which cost the life of Capt. Thomas Mantell of the Kentucky Air National Guard.

The two incidents actually have a lot in common. The protagonist in the dogfight, Lt. George Gorman, was a decorated and experienced World War II fighter who continued to fly in the Air National Guard, just like Capt. Mantell. The incidents also involved the exact fighter plane, the legendary P-51 Mustang, which won control of the skies over Europe for the Allies in World War II. A key difference was that Gorman survived, while Mantell was killed.

From Night Flight to Dogfight

The "dogfight" began on Oct. 1, 1948, when Gorman and other National Guard pilots were on a cross-country training flight in Mustangs. The flight arrived over Fargo, N.D. around 8:30 p.m. The other pilots decided to land for dinner, while Gorman decided to get in some night flying.

Around 9 p.m., Gorman encountered two objects in the sky, a Piper Cub civilian plane and a mysterious blinking light. Gorman decided to investigate the mysterious blinking object. When he passed by the object, Gorman saw it to be a ball of light, which he said was around six to eight inches in diameter.

Gorman flew at the object, which tried to avoid him by making a 180-degree turn, then flying straight at him. The object dived steeply and behaved like a piloted plane trying to avoid an enemy in combat; hence the use of the term "dogfight" to describe the encounter. Gorman and the object pursued each other for about half an hour before the light broke off and flew away.

A Strange Encounter

The dogfight was unusual because it involved a military plane and a highly experienced pilot, a World War II veteran. Gorman had flown in the Navy in the Pacific and he had apparently seen combat with the Japanese.

There were also witnesses, including Dr. A.D. Cannon, the pilot of the Piper Cub, and air traffic controllers in the control tower at Hector Airport in Fargo. These people backed Gorman's account of the events and his claims that the object behaved as if it was being directed by thought.

Gorman swore to investigators that he thought the object was piloted. A few hours after the dogfight, Gorman and the witnesses were interrogated by investigators from Project Sign, an Air Force enquiry into UFOs. Intriguingly, the investigators examined Gorman's Mustang with a Geiger counter and discovered that the plane was mildly radioactive.

Controversy and a Weather Balloon

The Gorman case attracted national attention and sparked controversy when the Air Force concluded that what the pilot had seen was a weather balloon. The investigators dismissed the maneuvers Gorman and the witnesses described as optical illusions.

The National Weather Service had released a balloon in the area, but this claim seems improbable. The object Gorman described was smaller than a balloon, and a balloon cannot maneuver. The rest of the Air Force's explanations are also hard to swallow – the investigators claimed he had been chasing the planet Jupiter.

It is hard to imagine an experienced pilot mistaking the planet Jupiter for a piloted craft. It is clear that Gorman saw something more than a weather balloon or the planet Jupiter, even though some UFO researchers accept the Air Force's story. A later investigation claimed that Gorman might have been chasing a Royal Canadian Air Force jet fighter. That sounds improbable because the Canadian pilot would have radioed Gorman and the control tower.

When these answers failed to hold water, the Air Force came up with an even more bizarre "explanation" that Gorman had been chasing a reflection of his own plane. These explanations indicate that the Air Force had no adequate explanation for the dogfight or at least one that they wanted to reveal to the public.

What Did Gorman See?

There are some factors that add credibility to Gorman's claims, including the other witnesses and the size of the object – it was only a few inches in diameter. That doesn't sound like a standard UFO sighting or something people would make up.

So what did Gorman fight, and what did the witnesses see? An intriguing possibility is that he encountered some sort of drone or probe similar to those sent out by NASA. It's interesting that the probe tested the plane's abilities by engaging in a dogfight with it. One possibility that some suggest is that aliens were attempting to ascertain Earth's military capabilities.

At the time, the United States was the world's greatest military power and the only nation to possess nuclear weapons. Other incidents of the era, such as the Mantrell encounter and the Roswell incident, took place near U.S. military installations.

There is one other intriguing fact about the Gorman Dogfight. Lt. Gorman was prevented from discussing the incident with the public by threat of court martial and dismal from the service. The question remains that if he had seen nothing but a reflection, then why did the Air Force go to such lengths to try and silence him?

The Great Los Angeles Air Raid

The largest and most public UFO sighting in United States history is also one of the least known. The Great Los Angeles Air Raid, or the Battle of Los Angeles, occurred on Feb. 25, 1942. The sighting was well-documented and covered extensively in the media, but it wasn't recognized as a UFO encounter at the time.

City in Terror

The incident is called the Battle of Los Angeles because artillery pieces on the ground actually opened fire on a giant object passing overhead. Citizens panicked when they saw this happening because it was a little over two months after the Japanese attack on Pearl Harbor. America had just entered World War II and, so far, the U.S. was losing the war.

Many Americans assumed that California would be bombed or invaded by the Japanese. Others were afraid of sabotage and espionage on the part of Japanese Americans. The situation in Los Angeles in 1942 was so tense that most citizens welcomed the federal government's illegal and unconstitutional roundup of Japanese Americans.

It was against this backdrop that one of the biggest UFO incidents in history occurred, yet almost nobody noticed that a UFO may have been present because of the ongoing war.

The Air Raid that Wasn't

The Battle of Los Angeles began around 2:25 a.m. on Feb. 25, 1942. The U.S. Army's Western Defense Command suddenly ordered the cities of Los Angeles and San Diego and their surrounding areas blacked out. A blackout was a precaution against an aerial attack, and the idea was that bombers wouldn't be able to find their targets if there was no light on the ground.

Shortly after the blackout began, antiaircraft artillery around aircraft plants in Los Angeles, Santa Monica, and Long Beach started firing. It wasn't clear what the guns were firing at, but it definitely was not Japanese planes.

People who went outside to look saw unusual objects flying through the skies. They noticed something else unusual – the gunfire was having absolutely no effect on the objects in the sky. Witnesses said the craft in the air seemed impervious to artillery fire. A reporter for *The Los Angeles Herald-Express* newspapers said a craft he saw was hit by dozens of shells, which had absolutely no effect upon it.

UFOs Flying in Formation

Witnesses on the ground were able to see what the artillery was firing at – bright red objects moving in formation. Those on the ground described the objects as blobs of light.

Newspaper photographs showed the objects, some of which were caught in searchlights. During World War II, giant searchlights were sometimes used to search the sky for bombers. The searchlights apparently lit up some of the UFOs. A famous picture from *The Los Angeles Times* shows searchlights trained on a giant UFO.

Witnesses described a giant craft the size of a blimp or a dirigible flying over the city. People were confused because the Japanese had no airships. The U.S. Navy did have blimps, but a blimp would have been easily shot down by artillery. Eyewitness reports make clear that whatever was hovering over Los Angeles on Feb. 25, 1942 was larger than any known aircraft.

The giant object hovered over the city for about 30 minutes. It moved down the coast from Santa Monica to Long Beach, then vanished as mysteriously as it appeared.

Aftermath

The UFOs apparently stayed over the city for nearly an hour, but the artillery kept firing for nearly two hours. On the ground, air raid wardens reported to their posts and citizens prepared for bombs that never came.

There were some casualties from the battle – at least two people on the ground were hurt by debris from falling shells. Some dairy cattle were reportedly killed by shells, and a Long Beach police officer was killed in a traffic accident during the fracas.

Yet when the "battle" ended, there were no bombs and no signs of enemy activity. Los Angeles tried to get back to normal, and military officials began scrambling to cover up the incident.

Cover-up

The Battle of Los Angeles led to what may have been the first official UFO cover-up in history. Authorities didn't want to admit that there were flying craft they couldn't shoot down. They may have also been afraid of panicking the public because just two days earlier, a Japanese submarine had shelled the area north of Santa Barbara.

When Los Angeles newspapers demanded an official explanation, what they got in response was doubletalk. The Secretary of the Navy, Frank Knox, blamed the incident on war jitters. Knox, a former newspaperman, had a habit of lying to the press to make himself look good.

The military wanted to keep the matter quiet because there had been quite a bit of bungling. The artillery had fired, but not hit anything, and some of the artillery fire had been delayed. To make matters worse, no fighter planes had been scrambled, even though there were some available in the area.

Unfortunately, no official investigation was made of the battle because of the pressing demands of World War II. The only official response was to round up everybody of Japanese descent and place them in what President Franklin D. Roosevelt admitted were concentration camps. The reason for this was to prevent sabotage, even though the federal government knew there no saboteurs. The fear was that Japanese Americans were in contact with the aircraft, another claim that federal authorities knew to be false and was used as a pretext.

Some conspiracy theorists have suggested that "the battle" was staged to provide a pretext for the internment of Japanese Americans, but no corroborating evidence for this theory has ever been found. Another popular theory was that military authorities wanted to move aircraft plants inland where they would presumably be safer from Japanese attack. Those cynics believe "the battle" was staged to provide a reason to move the factories, but the factories were never moved.

The Unsolved Mystery

So what did the artillery fire on during the Battle of Los Angeles? The truth is that nobody knows, although descriptions from witnesses note that the objects in the sky sound a lot like modern UFOs. A number of giant dirigible-shaped UFOs were seen right after the war.

The Battle of Los Angeles was quickly forgotten amidst the excitement and hysteria of the war. Modern UFO experts didn't start looking into the affair until the 1980s and haven't been able to determine what happened. Yet it is possible that the U.S. military fired on a number of UFOs over America's second-largest city during World War II and nobody realized what had happened for nearly two generations.

The Mantell UFO Incident

Captain Thomas Mantell of the Kentucky Air National Guard may have been the first person killed by a UFO. Mantell died when his plane crashed while chasing a UFO in 1948.

Capt. Mantell's tragic death is of interest to UFO experts for a number of reasons. It was one of the incidents that launched the first wave of popular interest in unidentified flying objects; it occurred just after the Roswell Incident. It also involved the military and some mysterious circumstances.

The Hero and the UFO

Capt. Thomas F. Mantell Jr. was a highly experienced fighter pilot and a war hero. He had flown P-51 Mustangs against the Nazi Luftwaffe in Europe during World War II and had been awarded the Distinguished Flying Cross.

After the war Mantell returned home to Kentucky and joined the newly formed Kentucky Air National Guard. He got to fly Mustangs again and lead a fighter squadron. The circumstances of his death are unusual and highly suspect for a number of reasons.

The incident began on Jan. 7, 1948, when a large number of people in the area of Fort Knox spotted several large circular objects flying through the air. The objects reportedly flew close to Fort Knox, where the United States gold reserve is kept. The Fort also houses an important army base.

Interestingly enough, several members of the military, including the commander of Goldman Airfield at Fort Knox, Colonel Guy Hix, reported seeing an umbrella-shaped object that sat in the air for one and a half hours. Not knowing what the object was, Hix called on the nearest available flight of planes for help.

Chasing the UFO

Capt. Mantell was in the air leading a training flight of four P-51 Mustangs to Strandford Air Field. Three of the Mustangs changed course and moved to intercept the object near Fort Knox.

When the planes approached the object, the pilots realized it was sitting at 30,000 feet, which was higher than the Mustang was supposed to fly. The Mustang was one of the fastest propeller-driven fighters ever built, but it lacked a pressurized cabin, so it couldn't climb to such altitudes without killing the pilot. Only one of the three pilots involved, Lt. Albert Clemmons, had an oxygen mask.

Even though the object was out of range, the three pilots tried to reach it. Two of them, including Clemmons, who had run out of oxygen, turned back, but Mantell kept up the pursuit.

What Did the Captain See?

An experienced pilot like Mantell knew it was dangerous and perhaps certain death to try and climb to above 20,000 feet, but he did so anyway. The question experts have been asking for over 65 years is why? What did the Captain see that compelled him to push his body beyond the levels of endurance?

Another question is why did he try to climb that high? Was Mantell perhaps under orders to try and get a good look at the UFO? The Air Force was interested in UFOs; at the time, it had a formal effort to investigate them known as Project Sign.

We'll never know the answer because, at some point, Mantell blacked out from lack of oxygen, which caused his plane to crash. At least that's the most likely explanation. The plane crashed, and Mantell didn't parachute out. His dead body was found in the plane on the ground.

No Explanation

The Mantell incident still attracts interest because it is still listed by the Air Force as unsolved. The Air Force gave no official explanation for the crash or the reason Mantell was pushing himself so hard.

Newspaper reports at the time of the crash gave other more questionable explanations for his death that raise eyebrows. One report suggested that radiation or a magnetic ray fired by the UFO caused Mantell to crash or killed him in the air.

Some of the reports indicate that the mysterious object moved to a higher altitude when the fighter planes approached it. That sounds like the behavior of a manned aircraft.

Even though the Air Force didn't provide an official explanation, the press was later told that Mantell had been chasing a "weather balloon." Unmanned balloons don't suddenly change altitude when aircraft approach.

So What Was It?

We'll probably never know what killed Capt. Mantell, although there are some possible explanations. Mantell may have seen an experimental balloon launched by the U.S. government; such experiments, called Skyhook, were ongoing in 1948. He may have been under orders to shoot the Skyhook balloon down.

Another possibility is that Mantell saw a Soviet balloon of some sort. It was 1948, and the Cold War was beginning to heat up. The balloon was floating around an important U.S. military installation. Balloons with cameras in them were used for observation before the invention of satellite surveillance.

Or he might have seen a real flying saucer. Such craft were in the news at the time, and the Air Force and Air Guard wanted to learn about them, particularly one flying around an important military installation such as Fort Knox.

The truth will never be known because of Capt. Mantell's tragic death. Mantell's body was recovered, and he was buried at the Zachary Taylor National Cemetery in Louisville, Kentucky. His sacrifice has not been forgotten; in 2001 the Simpson County Kentucky Historical Society erected a marker honoring his memory.

The Nash-Fortenberry UFO Sighting

Most UFO sightings involve just one object, but on July 14, 1952, airline pilots William Nash and William H. Fortenberry saw what appeared to be a squadron of what they called "flying saucers" flying in formation. Interestingly enough, the pilots flew over the objects rather than under them.

The sighting occurred when Nash and Fortenberry were piloting a Pan American DC-4 from New York City to Miami. The plane was over Chesapeake Bay not far from Newport News, Va., one of America's most important Navy bases, when the unidentified flying objects were sighted.

The two first noticed what they described as a brilliant red light in the sky. When they looked closer, the two saw six objects moving at a very high speed. All of the objects were red-orange in color and circular. The UFOs were moving in formation like fighter planes.

UFOs Flying in Formation

Nash and Fortenberry reported that the objects behaved much like fighter planes. They maintained a tight formation and changed speed to keep up with each other. The objects were saucer or coin-shaped and were travelling about 2,000 feet about the ground.

The objects changed direction sharply and maneuvered like airplanes might. Yet they were clearly not planes because the two estimated their speed at 1,000 miles per hour. The pilots also noted that the objects made a 150-degree turn; something clearly impossible for even a jet fighter in the early 1950s. When the objects sped up, the pilots estimated the speed at 12,000 miles per hour.

Fortenberry and Nash both clearly believed that what they saw was the product of an alien intelligence. In a 1967 article for *True Magazine,* the two noted that no material known to man could stand up to the speeds the objects were moving at. They also noted that human beings could not have stood up to the speeds the objects were moving at either.

A Credible Sighting

The Nash and Fortenberry sighting is considered very credible because the two men were both former U.S. Navy pilots who had been trained in enemy aircraft identification during World War II. Intriguingly, the two were interviewed by five U.S. Air Force intelligence officers when they arrived in Miami the next morning.

The high level of military interest indicates that the Air Force had knowledge of the sighting from somewhere else. At least one member of the military, a Navy officer serving on the Cruiser U.S.S. Nashville, wrote he had sighted eight red lights moving across the sky on the night of July 14 in a letter to the editor of *The Norfolk Virginian-Pilot* newspaper.

There are some holes in Nash and Fortenberry's story. They provided two different counts of the number of objects they saw: eight and six. The third pilot on the plane, Captain F.V. Koepke, the craft's commanding officer, did not corroborate their story nor did any of the passengers on the DC-4 report seeing the objects.

What Did They See?

The Nash-Fortenberry Sighting is intriguing because two veteran pilots trained in identification observed a number of objects displaying evidence that they were being directed by some sort of intelligence. The consistencies in their story add credibility, as does the fact that they would have little to gain by making up such a tale.

The two obviously saw something, although their description of the objects conflicts with the UFOs seen in the Sperry and Chiles-Whitted encounters in the same era, which also involved airline pilots. The pilots in those two incidents reported seeing giant cigar-shaped objects moving at a high speed, not saucers.

It is interesting to note that the encounter occurred near a major military installation: the naval base at Newport News, the headquarters of the Atlantic Fleet. Aircraft carriers are based at Newport News and so are some of the Navy's jet fighters. Other UFO incidents of the era, including the Mantell incident, also occurred near military bases.

The Air Force reported that five jets were in the air on the night of July 14, 1952. Yet the objects Nash and Fortenberry saw were clearly flying at speeds no jet ever built is capable of reaching. A jet would have disintegrated at the speeds the two pilots reported the objects moving at.

Certain That They Saw Aliens

The most interesting thing about the Nash and Fortenberry account is that the two were certain that they had glimpsed an alien intelligence. Even 15 years later they stood by their claim in a magazine account. Although the claims might have damaged their careers, the two were convinced that they had seen real alien craft flying through the air.

So what did the two see? Perhaps they glimpsed smaller craft released from some sort of mother ship. Whatever they saw, Nash and Fortenberry saw objects demonstrate capabilities that are still well beyond the limits of human technology.

Valentich Disappearance

Encounters between UFOs and aircraft usually don't lead to disappearances, which is what makes the incident involving Australian Frederick Valentich so fascinating. Both the young man and the plane he was piloting vanished completely.

Valentich's disappearance is of interest to UFO enthusiasts because he claimed a mysterious aircraft was orbiting above him in his last known radio conversation with the control tower. Neither Valentich nor the plane he was flying were ever seen again after the mysterious transmission.

To add to the mystery, a number of people reported UFO sightings in the area where Valentich disappeared. No satisfactory explanation for Valentich's disappearance has ever been found, and neither his plane nor his body has ever been recovered.

The Mysterious Flight to Nowhere

Frederick Valentich was a young resident of Melbourne, Australia who was studying to be an airline pilot in 1978. He was also an amateur pilot with some experience, and he had access to a private plane, a Cessna 182-L.

Valentich left Moorabbin Airport in Melbourne around 6 p.m. in the evening. Before he took off, Valentich had filed a flight plan that indicated he was flying to King Island off the coast to pick up passengers. The flight plan later raised questions because Valentich had told his family and his girlfriend that he was going to the island to pick up crayfish.

Accident investigators later learned that there were no passengers waiting for Valentich and no crayfish for him to pick up. This is one of several mysteries associated with the flight that has never been explained.

It's Not an Aircraft

Frederick Valentich was last heard around 7:09 p.m. on Oct. 21, 1978 when he radioed the control tower in Melbourne. It's this message more than anything else that has raised the claims that Valentich was abducted by aliens aboard a UFO.

Valentich told the controller that a giant aircraft was orbiting above him. He described the craft as having a shiny metal surface and a green light. He then said the aircraft had vanished, but soon reported that the craft was approaching from the Southwest.

Valentich next reported engine trouble and told the controller, Steve Robey, "It is hovering and it's not an aircraft." That remark was followed by 17 seconds of mysterious scraping sounds after which all contact with him was lost.

Search and UFO Reports

When Valentich didn't land on King Island as scheduled, a search was immediately launched. Two planes from the Royal Australian Air Force searched the area of ocean where he had last reported and found no traces of him and his plane.

An oil slick was spotted, but when it was tested, searchers determined that the substance in the water was not aviation gas. Reports didn't say what this substance was or if it was related to Valentich's disappearance.

After the disappearance, several people came forward with UFO reports, including some people that saw a mysterious green light in the sky near Apollo Bay where Valentich last radioed. Some of the witnesses claimed that the green light was following Valentich's plane, and another said that the object was several times larger than Valentich's plane.

A man named Roy Manifold came forward with some unusual pictures taken around the time of Valentich's disappearance. Manifold had set up an automated camera to take pictures of the sunset from the beach. The camera also took pictures of a fast-moving object in the sky. UFO experts examined the photographs and concluded Manifold had taken a picture of a UFO.

No Explanation

A number of investigations have been made into Frederick Valentich's disappearance and none of them have reached a satisfactory conclusion. An official investigation by the Australian Department of Transport was unable to determine the cause of the disappearance. Instead, the Department simply ruled that Valentich was presumed dead.

Investigations by various UFO groups couldn't reach a conclusion either. Analysis of the tape of the radio conversation did not reveal the cause of the mysterious sounds.

There has been pressure to mount an underwater search for the remains of the plane, but this has never been done. It isn't clear if the plane could be located 35 years after the incident.

Questions Remain

Questions remain about the disappearance and about Valentich himself. Valentich was reported to be something of a failure who had flunked a test for a commercial pilot's license five times. Valentich had also failed to get into the Royal Australian Air Force and achieve his lifelong dream of flying. Instead of flying, Valentich was working in a clerical job at an army storage facility in Melbourne.

He might have been dissatisfied with life and faked his disappearance or committed suicide. The disappearance and the mysterious radio message might have been his way of sparing his family the embarrassment of suicide. This is only conjecture, as no suicide note was ever found.

The truth is that we will probably never know what happened to Frederick Valentich and the plane he was flying. There is simply no way to know whether he was abducted by aliens or if his body and the missing Cessna are at the bottom of the ocean.